How to use this book

A sample page

INSTRUCTION
What your child needs to do for the activity.

TITLE
The page title describes the skill your child will learn in these pages.

FUN ILLUSTRATIONS
Specially drawn illustrations which are fun, interesting and drawn at the right pedagogical level for your child.

COLOURFUL BORDERS
The page borders make each page as attractive as possible to stimulate your child.

EXAMPLE
The first one is done for you so you can show your child exactly what to do.

LOTS OF PRACTICE
Two pages where your child can practise and repeat the same skill to master it.

STICKERS
Place a sticker on each page as your child finishes.

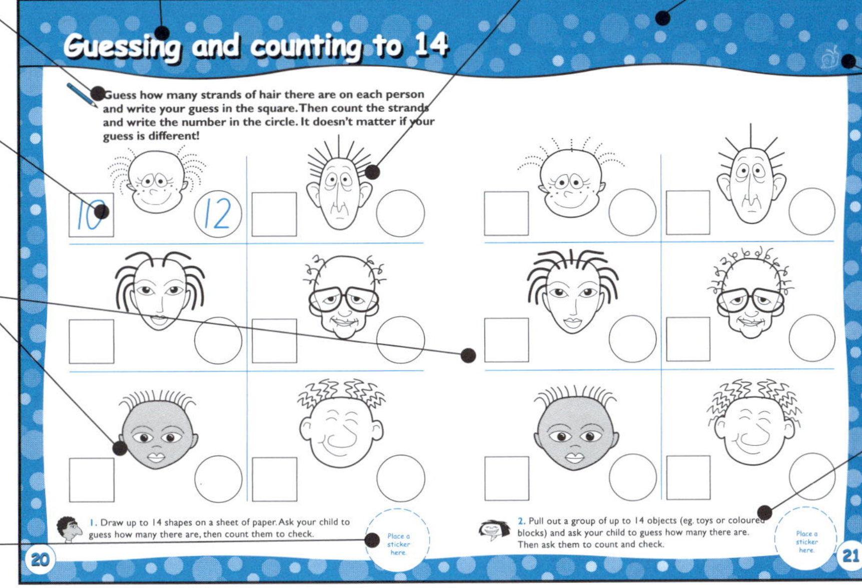

WHO'S HIDING?
In each book, a little creature appears in the border of every double page so your child can have fun trying to find it.

EXTRA ACTIVITIES
Extra activities you might want to do with your child to further reinforce the skill or simply make it more enjoyable.

Step-by-step learning

STEP ONE **Read** out the title of the activity page to your child.

STEP TWO **Explain** the skill and show your child the example already done. **Make sure** they understand what to do. Your child will then have at least two pages to practise that same skill.

STEP THREE **Help** your child put a **sticker** on the bottom of each page as they complete it.

Remember to be patient, encouraging and positive with your child, even when minor mistakes are made!

How to hold a pencil

It is important that you help your child hold his or her crayon or pencil in the correct way as shown here to ensure your child develops the right technique early on.

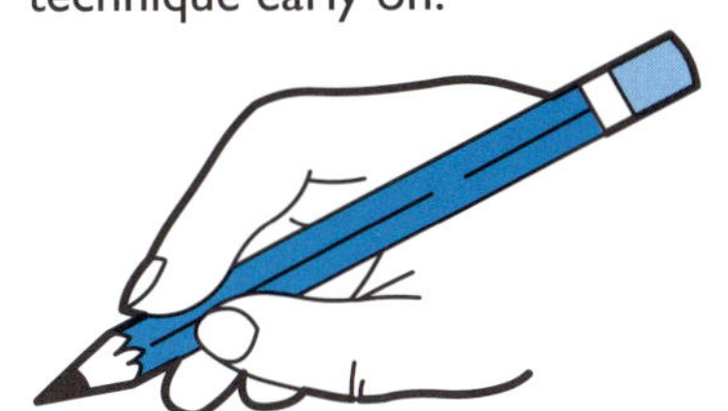

Reviewing numbers 1-9

Look at the cards hanging on the lines. Each line should have a number card in the middle, a matching shape card on the left and the name of the number on the right. Fill in the missing numbers, pictures or words.

5

five

nine

three

1. Help your child revise numbers 1 to 9 by counting objects around the home.

Place a sticker here.

seven

4

four

eight

2. Find ways to use these numbers when you are shopping. For example, ask your child to pick up a certain number of items from the shelves.

Place a sticker here.

Writing numbers before and

Write the numbers that come before and after the numbers on the middle shirt.

4 5 6

1 2

6 7

1. Draw a number line on a chalkboard or other surface, and leave some numbers out. Then ask your child to write in the missing numerals.

Place a sticker here.

after from 1-10

3 4

5 6

8 9

2. You could make some simple number cards up to 10 with one number on each. Jumble them up, then ask your child to put them in the correct order.

Place a sticker here.

Writing missing numbers fron

Write the missing numbers on each snake to show how you count from 1 to 9.

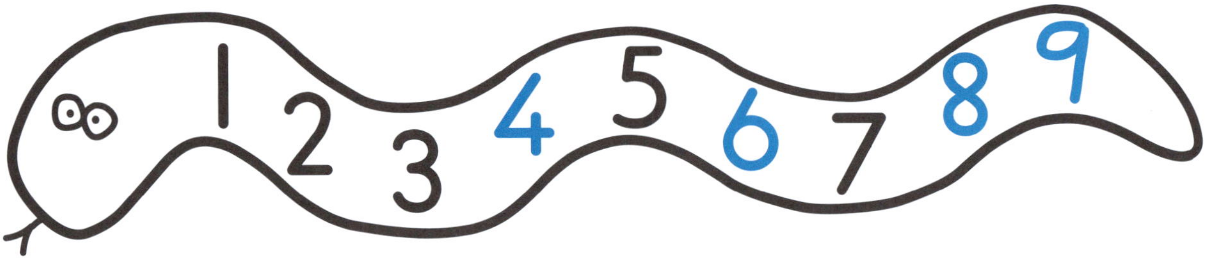

1 2 3 4 5 6 7 8 9

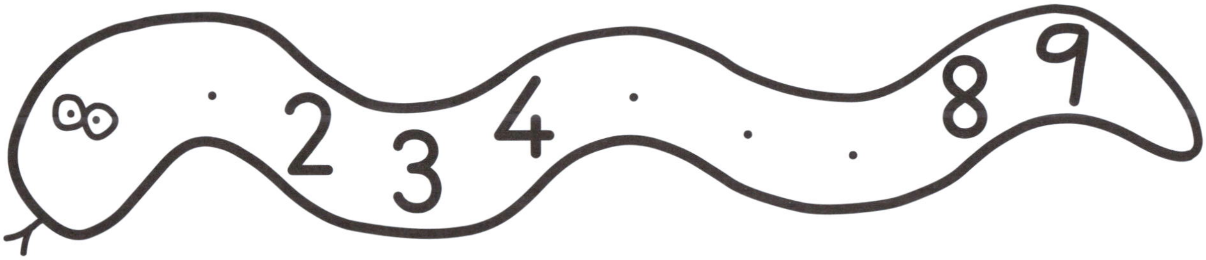

. 2 3 4 . . . 8 9

1 2 . 4 5 6 . . .

1 . . . 5 6 7 . 9

1. Count aloud as far as you can with your child past 10.

Place a sticker here.

1-10

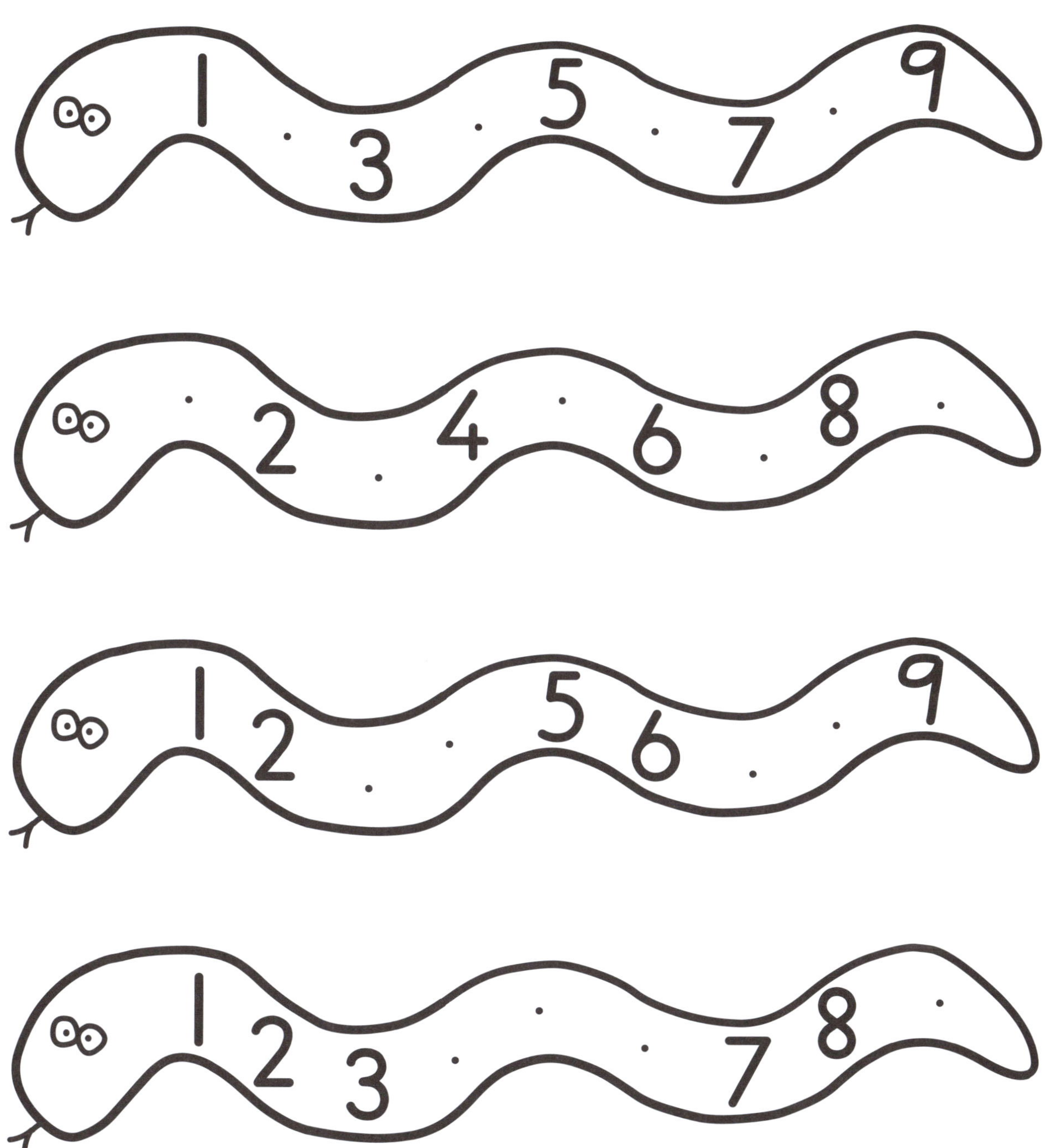

2. Ask your child to hold up 10 fingers and count them.

Place a sticker here.

Making groups of 10

There are many ways to make 10. You can hold up 10 fingers, or glue 10 dried beans on a stick. This gives you **1** group of 10 and no extras. See if you can make these groups of 10 things.

Trace 10 fingers.

1 group of 10 and 0 extra

Colour 10 petals.

☐ group of 10 and ☐ extra

1. Ask your child to hold up their hands and count '10 fingers and no extras'. This will help them later to count numbers greater than 10.

Place a sticker here.

Colour 10 frogs.

☐ *group of 10 and* ☐ *extra*

Draw 10 sausages.

☐ *group of 10 and* ☐ *extra*

2. How else can your child make a group of 10? 10 toys? 10 pieces of fruit in the basket?

Place a sticker here.

Making groups of 10 and 1

Making a group of 11 means making **1** group of 10 and **1** extra. See if you can make these groups of 11 things.

Trace 11 fingers.

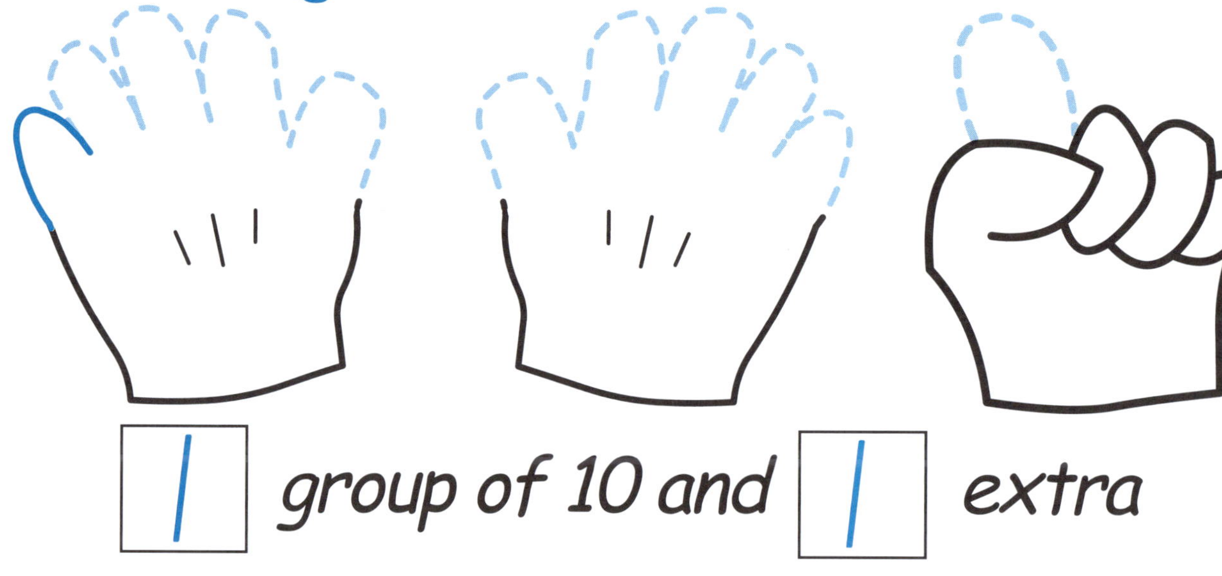

1 group of 10 and 1 extra

Colour 11 petals.

group of 10 and extra

1. Help your child make 1 group of 10 fingers and 1 extra. Then count them aloud to 11.

Place a sticker here.

Colour 11 frogs.

☐ *group of 10 and* ☐ *extra*

Draw 11 sausages.

☐ *group of 10 and* ☐ *extra*

2. Ask your child to cut out 11 pictures from a magazine and paste them onto a page. They could also make groups of 11 from objects around the home.

Place a sticker here.

Making groups of 10 and 2

Making a group of 12 means making 1 group of 10 and 2 extra. See if you can make these groups of 12 things.

Trace 12 fingers.

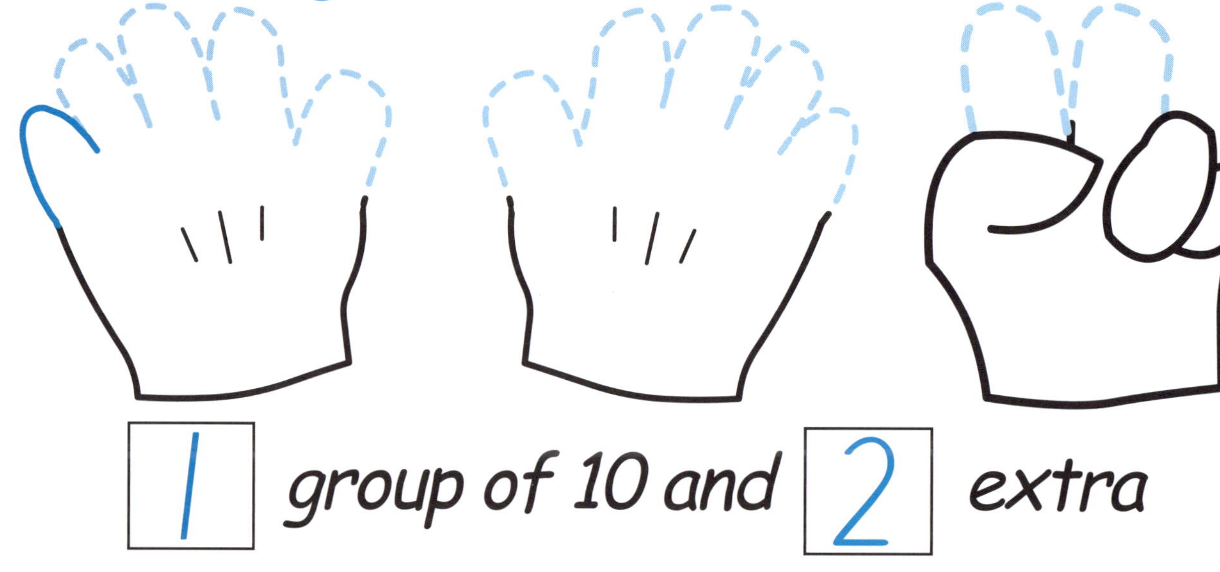

1 group of 10 and 2 extra

Colour 12 petals.

☐ group of 10 and ☐ extra

1. Clap your hands 12 times and have your child count the claps.

Place a sticker here.

extra

Colour 12 frogs.

☐ group of 10 and ☐ extra

Draw 12 sausages.

☐ group of 10 and ☐ extra

2. Ask your child to name 12 people you both know well.

Place a sticker here.

Guessing and counting to 1

Guess how many pictures there are in each group and write your guess in the square. Then count the things and write the number in the circle. It doesn't matter if your guess is different!

10

12

1. You can encourage your child to guess and then count groups of objects around the home.

Place a sticker here.

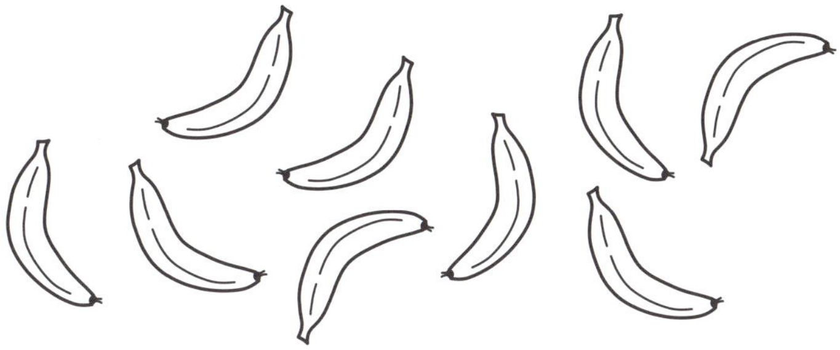

2. Ask your child to guess how many eggs are in a box of a dozen. Then they can count them to check.

Place a sticker here.

Making groups of 10 and 3

Making a group of 13 means making 1 group of 10 and 3 extra. See if you can make these groups of 13 things.

Trace 13 fingers.

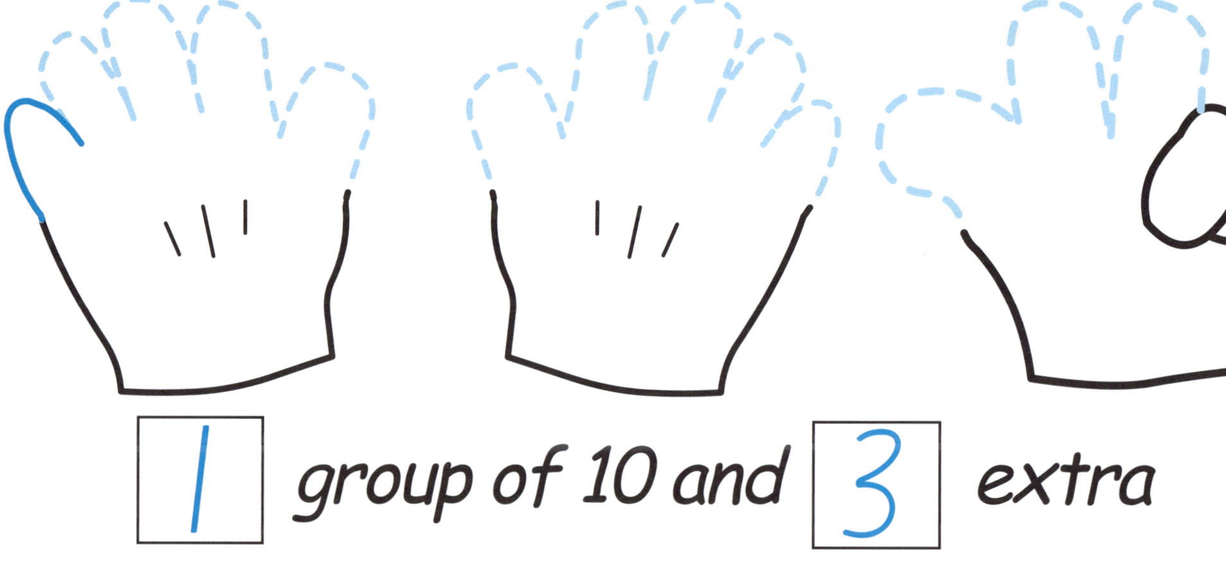

1 group of 10 and 3 extra

Colour 13 faces.

 group of 10 and  extra

1. Ask your child to jump up and down 13 times, counting aloud as they go.

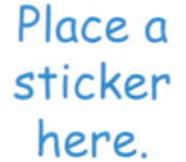

extra

Draw 13 spots on the cows.

☐ *group of 10 and* ☐ *extra*

Draw 13 olives on the pizzas.

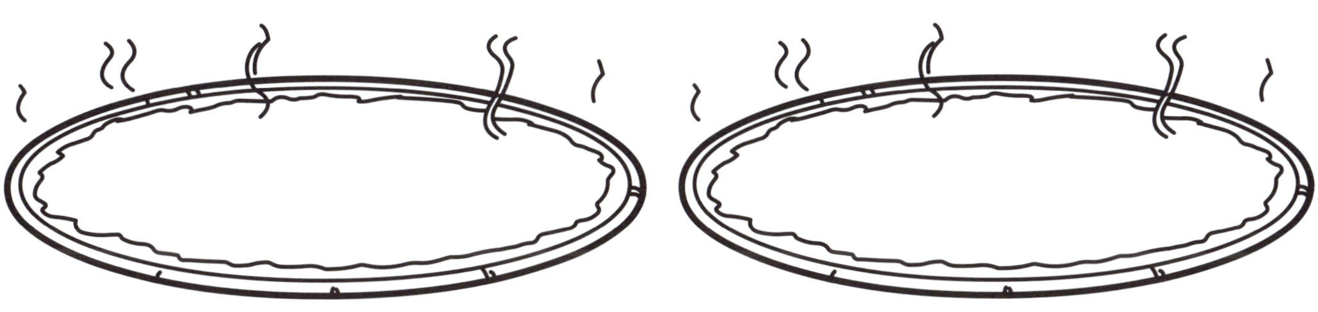

☐ *group of 10 and* ☐ *extra*

2. Clap 13 times while your child counts. Then ask your child to clap 13 times.

Place a sticker here.

Making groups of 10 and 4

Making a group of 14 means making 1 group of 10 and 4 extra. See if you can make these groups of 14 things.

Trace 14 fingers.

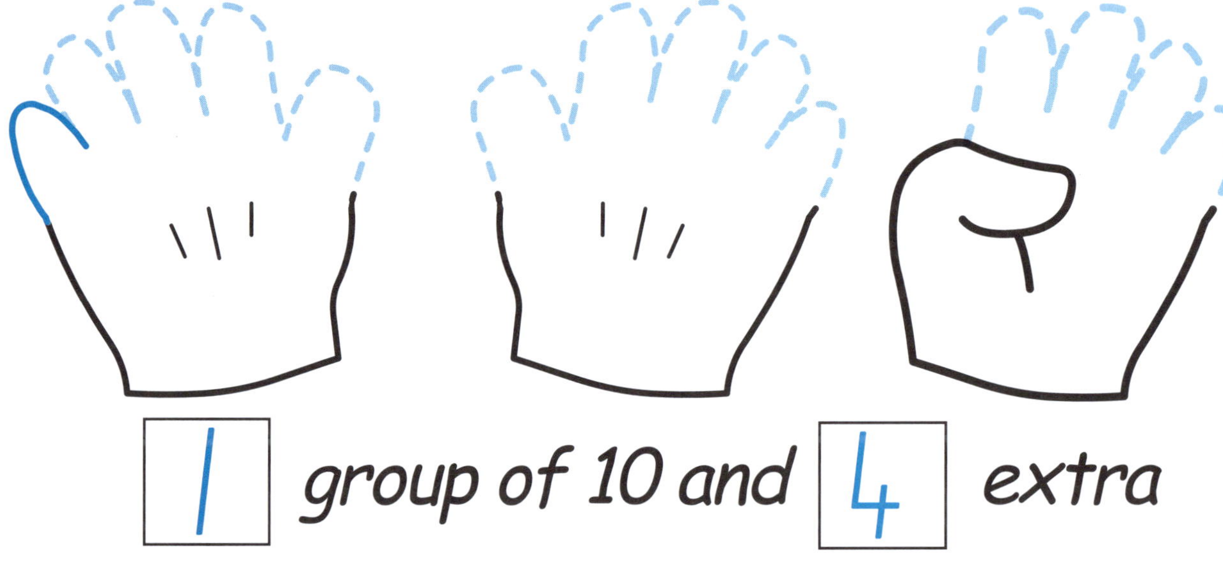

1 group of 10 and 4 extra

Draw 14 strands of hair.

group of 10 and extra

1. Ask your child to 'bend and stretch' 14 times, counting aloud on each one.

Place a sticker here.

Colour 14 beans.

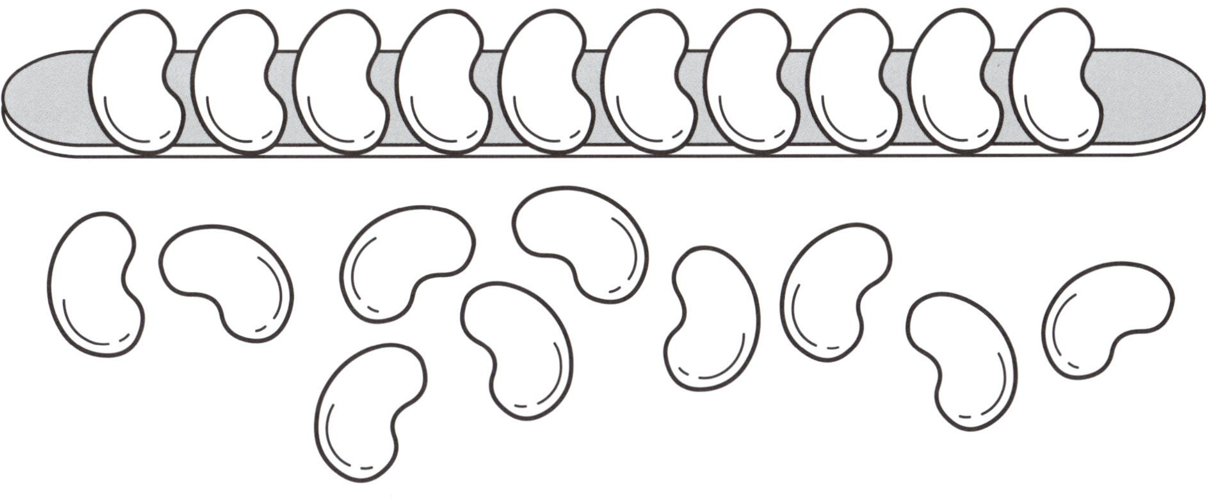

☐ group of 10 and ☐ extra

Draw 14 olives on the pizzas.

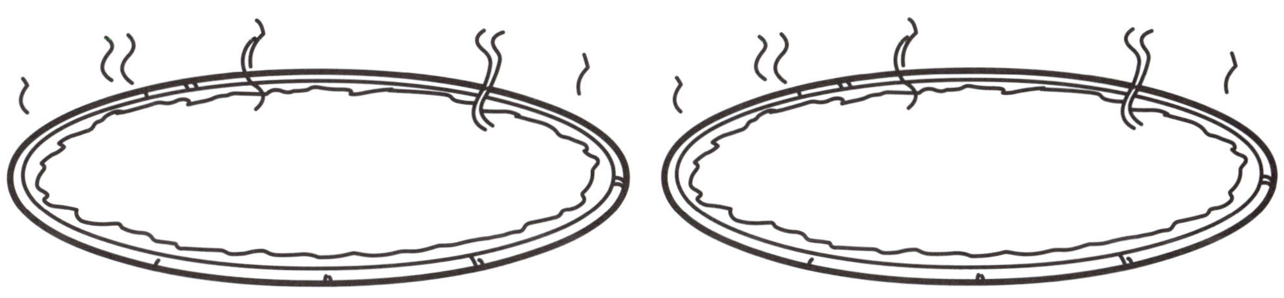

☐ group of 10 and ☐ extra

2. Ask your child to count to 14 with their eyes closed. Can they count backwards from 14?

Place a sticker here.

Guessing and counting to 1

Guess how many strands of hair there are on each person and write your guess in the square. Then count the strands and write the number in the circle. It doesn't matter if your guess is different!

10

12

1. Draw up to 14 shapes on a sheet of paper. Ask your child to guess how many there are, then count them to check.

Place a sticker here.

2. Pull out a group of up to 14 objects (eg. toys or coloured blocks) and ask your child to guess how many there are. Then ask them to count and check.

Place a sticker here.

Making groups of 10 and 5

Making a group of 15 means making 1 group of 10 and 5 extra. See if you can make these groups of 15 things.

Trace 15 fingers.

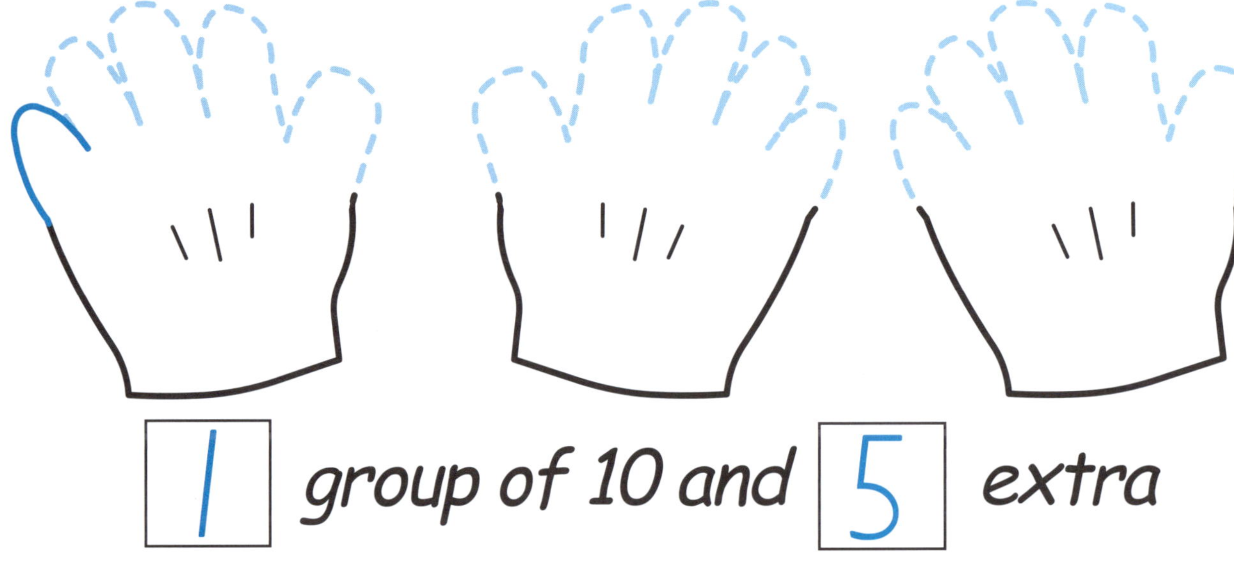

1 group of 10 and 5 extra

Colour 15 fish.

group of 10 and extra

1. Ask your child to flick their fingers 15 times, counting aloud on each one.

Place a sticker here.

Draw 15 apples.

☐ *group of 10 and* ☐ *extra*

Colour 15 bananas.

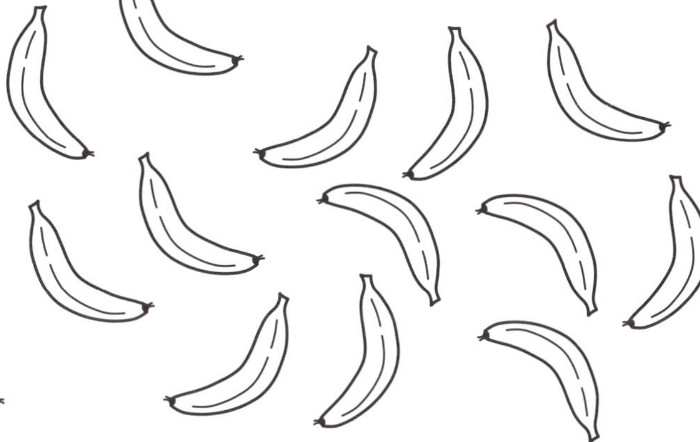

☐ *group of 10 and* ☐ *extra*

2. Ask your child to see how quickly they can count to 15. Then ask them to try it as slowly as they can.

Place a sticker here.

Making groups of 10 and 6

Making a group of 16 means making 1 group of 10 and 6 extra. See if you can make these groups of 16 things.

Trace 16 fingers.

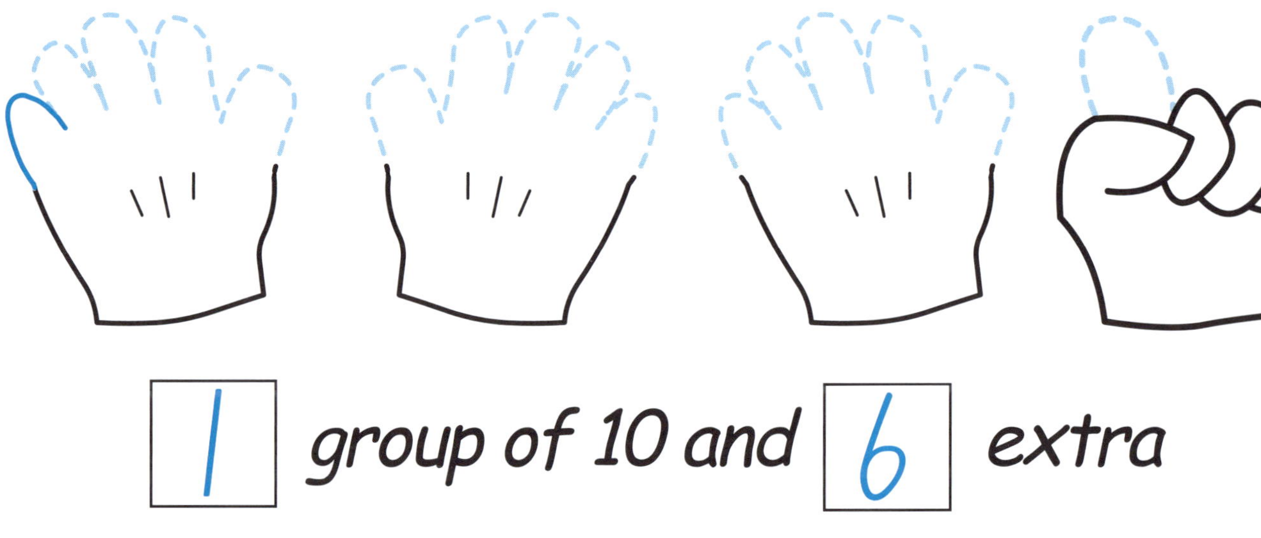

1 *group of 10 and* 6 *extra*

Draw 16 freckles.

☐ *group of 10 and* ☐ *extra*

1. Ask your child to nod their head 16 times, counting aloud on each one.

Place a sticker here.

Colour 16 beans.

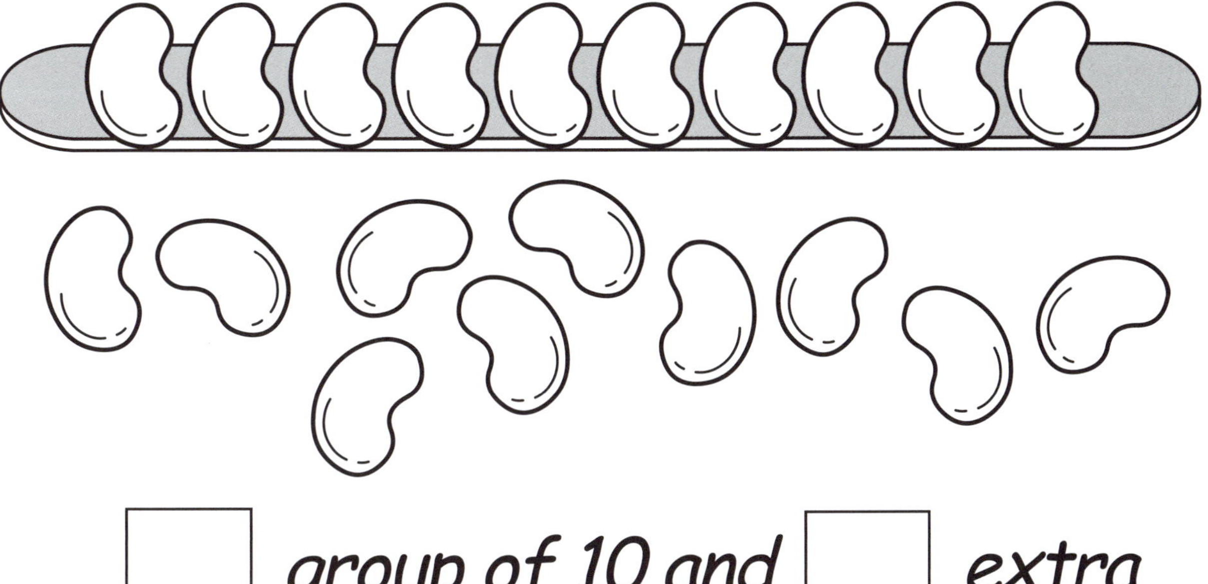

☐ *group of 10 and* ☐ *extra*

Draw 16 apples.

☐ *group of 10 and* ☐ *extra*

2. Give your child a group of 16 small items (eg. spoons). How many different ways can they arrange them into 2 groups (eg. 11 spoons + 5 spoons)?

Place a sticker here.

Guessing and counting to 16

Guess how many pictures there are in each group and write your guess in the square. Then count the things and write the number in the circle. It doesn't matter if your guess is different!

1. Ask your child to flap their arms 16 times, counting aloud on each one.

Place a sticker here.

2. Ask your child to collect 16 socks from their room.

Place a sticker here.

Making groups of 10 and 7

Making a group of 17 means making 1 group of 10 and 7 extra. See if you can make these groups of 17 things.

Trace 17 fingers.

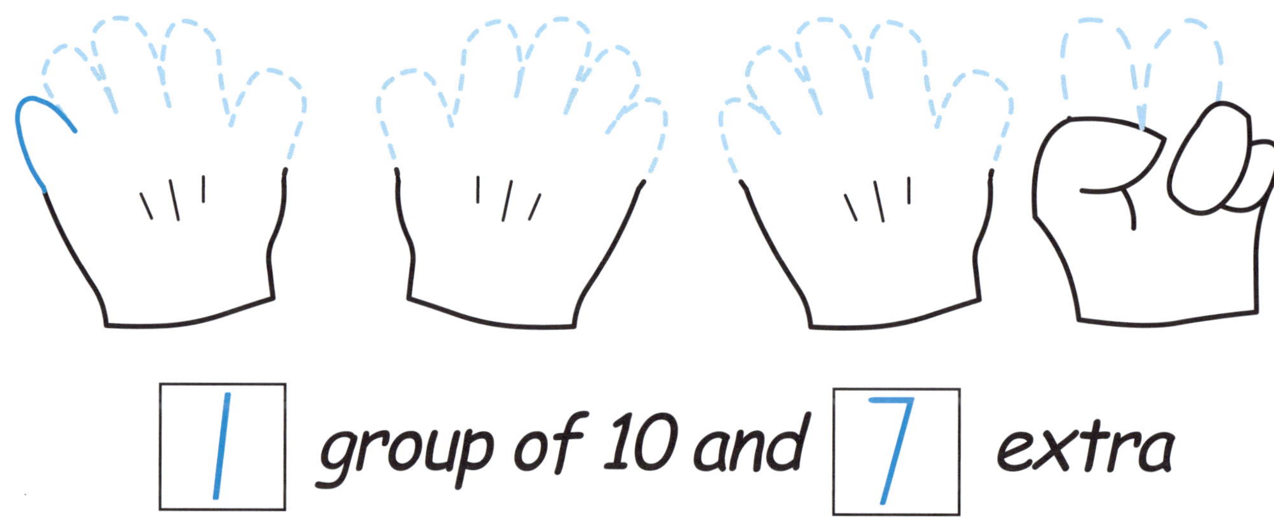

1 group of 10 and 7 extra

Draw 17 stripes.

group of 10 and  extra

1. Ask your child to shake their hands 17 times, counting aloud on each one.

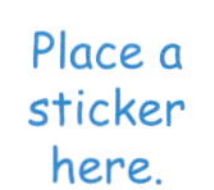

extra

Colour 17 beans.

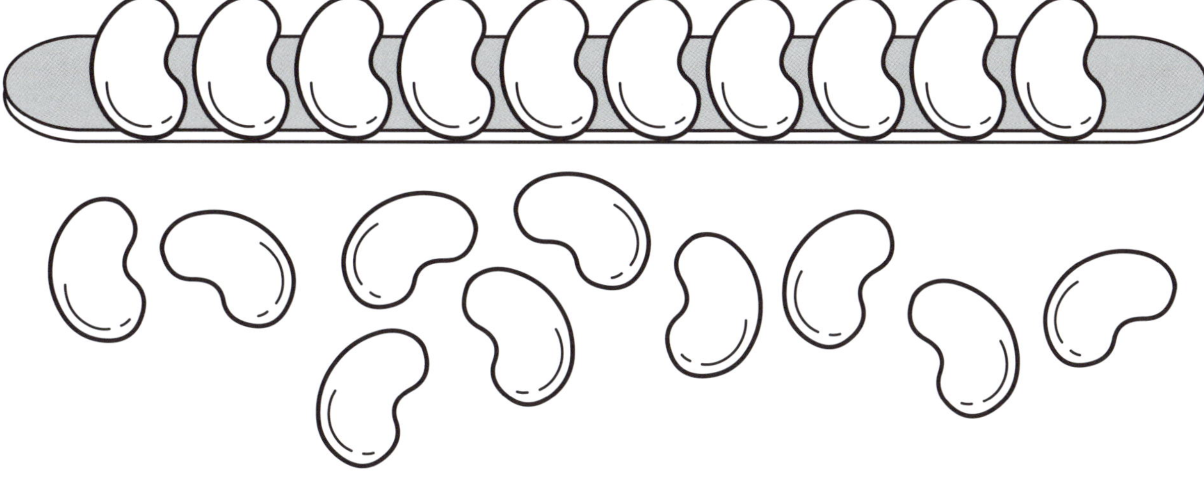

☐ *group of 10 and* ☐ *extra*

Colour 17 cans.

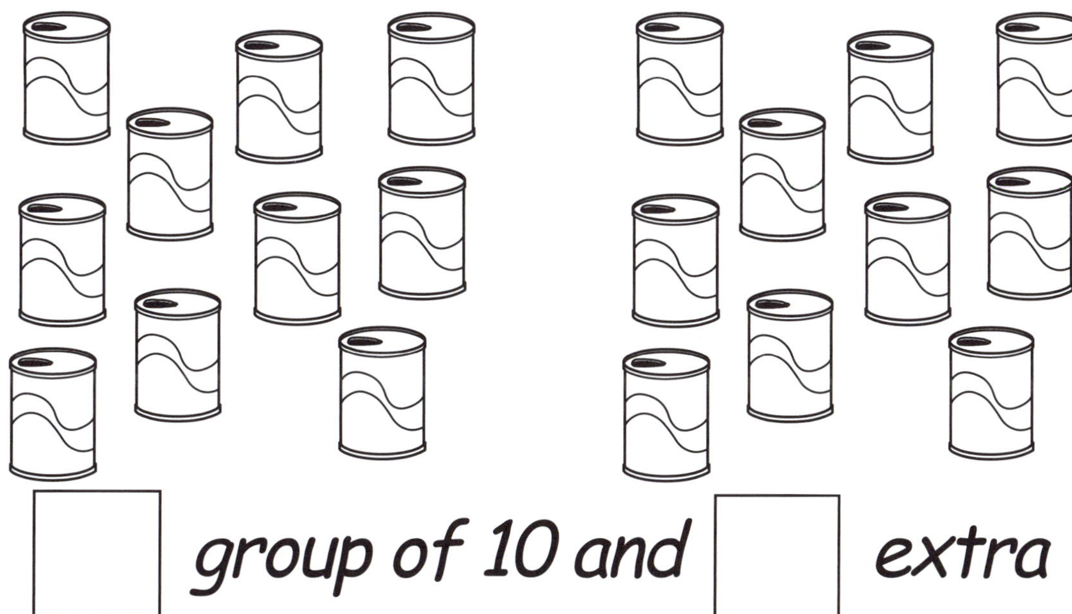

☐ *group of 10 and* ☐ *extra*

2. Flash between 10 and 17 fingers in front of your child for an instant. Can your child tell you how many fingers you held up? Try it again with a different number.

Place a sticker here.

Making groups of 10 and 8

Making a group of 18 means making 1 group of 10 and 8 extra. See if you can make these groups of 18 things.

Trace 18 fingers.

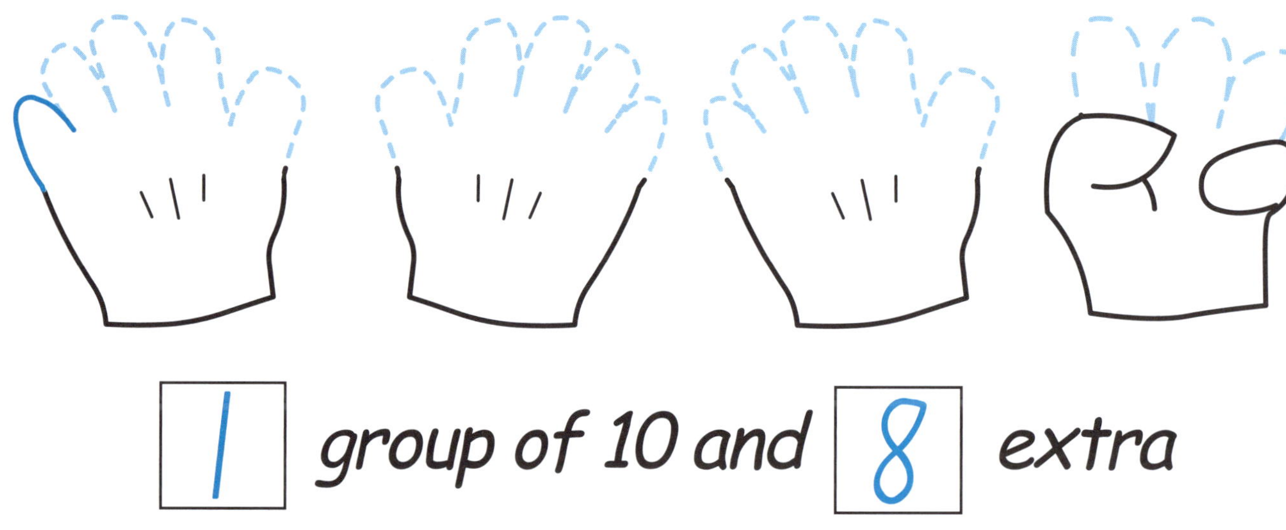

1 group of 10 and 8 extra

Colour 18 marbles.

group of 10 and ☐ extra

1. Ask your child to touch their toes 18 times, counting aloud on each one.

Place a sticker here.

Colour 18 cans.

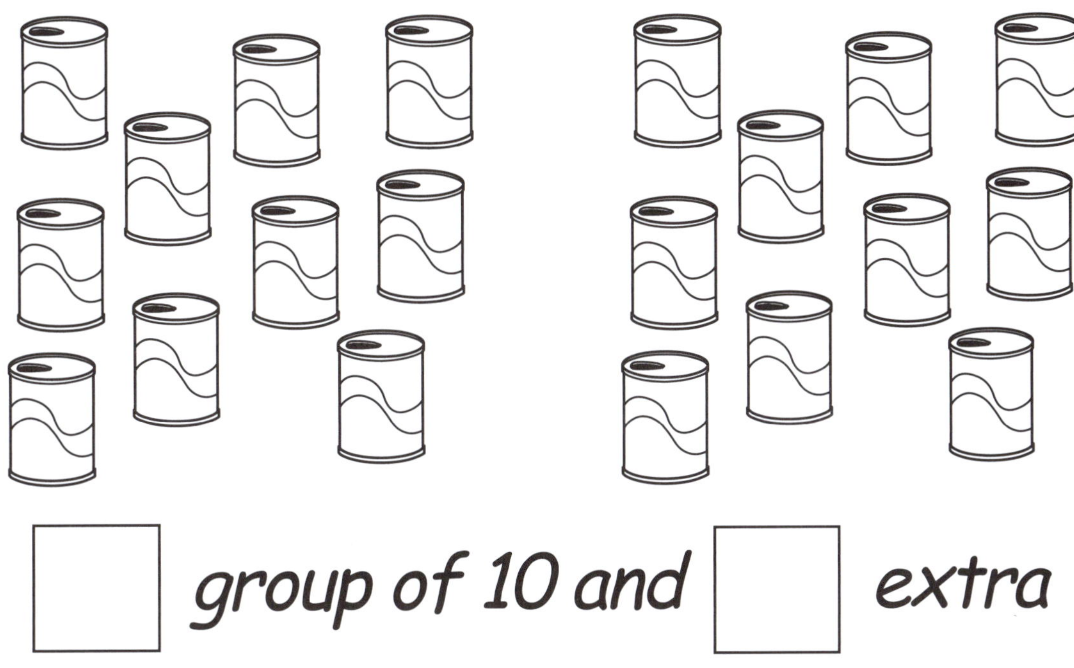

☐ *group of 10 and* ☐ *extra*

Draw 18 worms.

☐ *group of 10 and* ☐ *extra*

2. Give your child 18 ice-block sticks. How many different patterns can they make with them? You can stick them onto paper in your child's favourite pattern.

Place a sticker here.

Guessing and counting to 18

Guess how many dots there are altogether on each pair of cows and write your guess in the square. Then count the dots and write the number in the circle. It doesn't matter if your guess is different!

13

17

1. Show your child a handful of leaves up to 18. Ask your child to guess how many you have, then count them.

Place a sticker here.

2. Ask your child which is more — 18 spots or 15 spots? Help them count up from 10 to check.

Place a sticker here.

Making groups of 10 and 9

Making a group of 19 means making 1 group of 10 and 9 extra. See if you can make these groups of 19 things.

Trace 19 fingers.

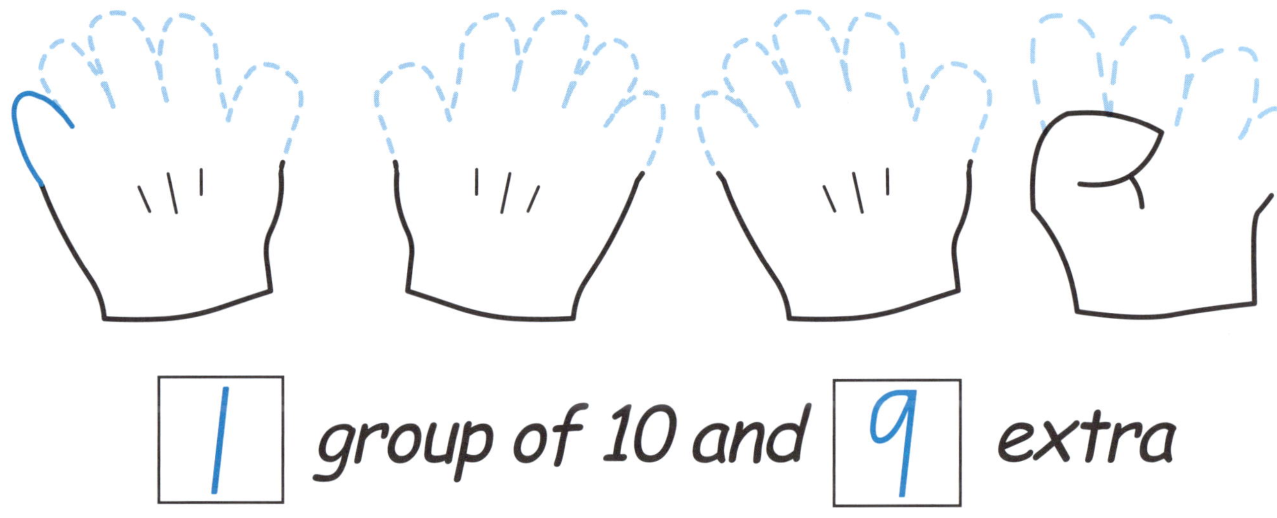

1 group of 10 and 9 extra

Draw 19 candles.

☐ group of 10 and ☐ extra

1. Ask your child to hop 19 times, counting aloud on each one.

Place a sticker here.

Colour 19 beans.

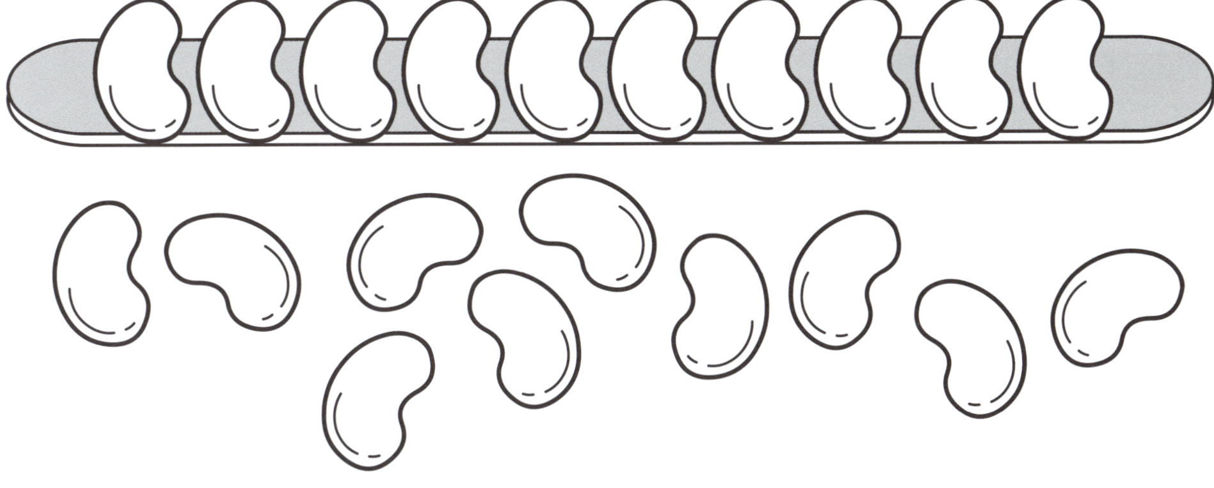

☐ *group of 10 and* ☐ *extra*

Colour 19 fish.

☐ *group of 10 and*  ☐ *extra*

2. Ask your child which is less — 17 fish or 19 fish? Help them count up from 10 to check.

Place a sticker here.

Making 2 groups of 10

Making a group of 20 means making 2 groups of 10 and no extras. See if you can make these groups of 20 things.

Trace 20 fingers.

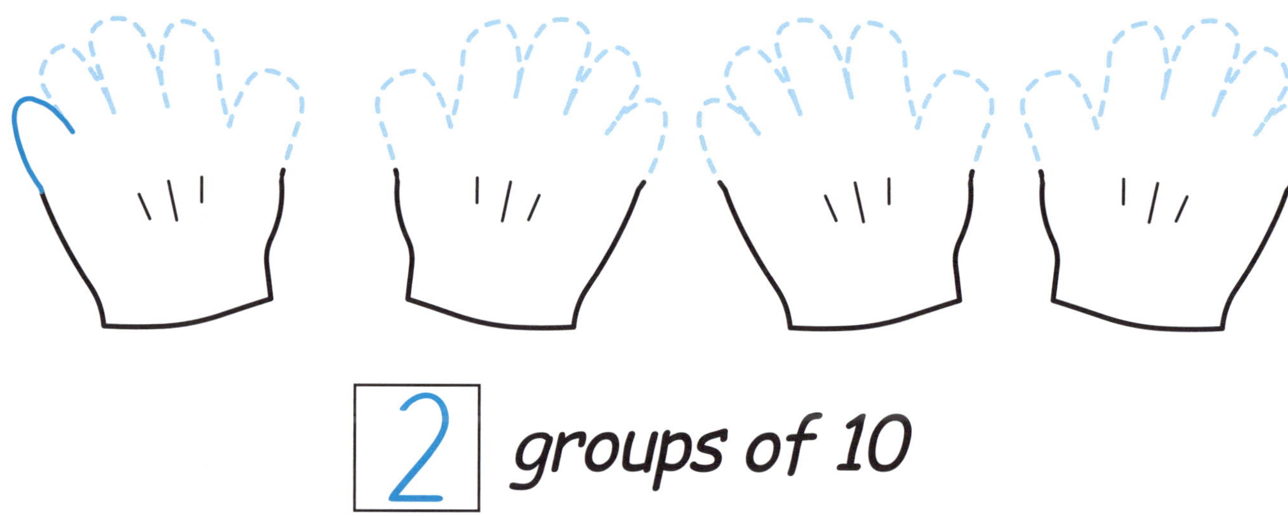

2 *groups of 10*

Draw 20 candles.

☐ *groups of 10*

1. Clap hands with your child 20 times, counting aloud on each one.

Place a sticker here.

Colour 20 beans.

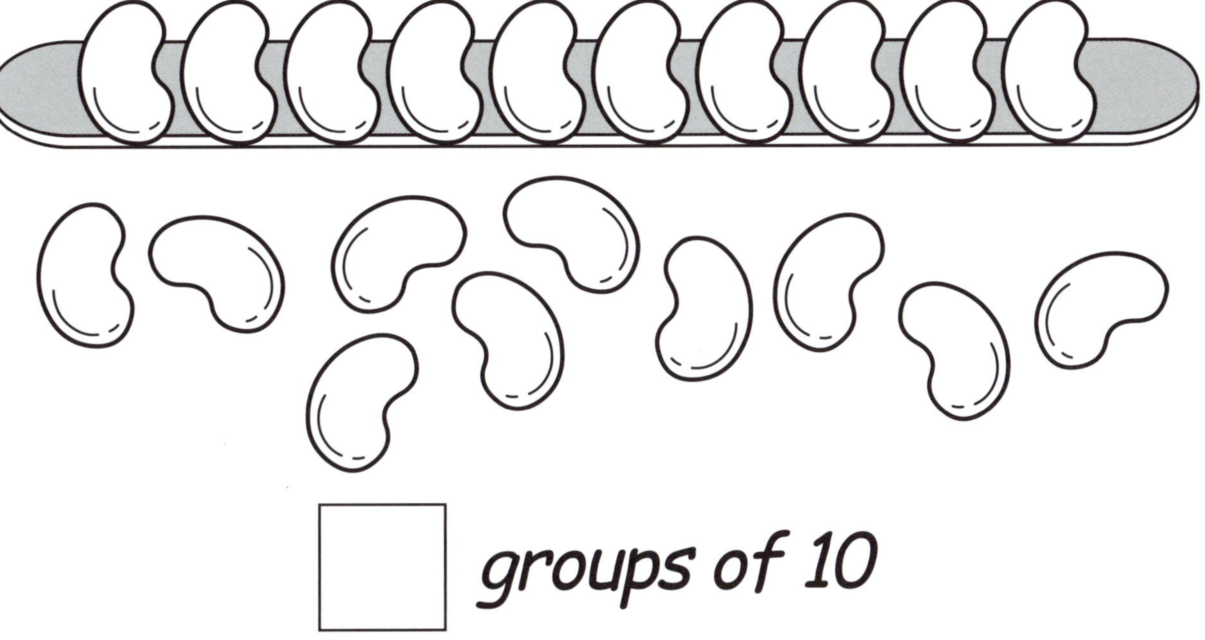

☐ *groups of 10*

Draw 20 strands of hair.

☐ *groups of 10*

2. Draw a large, simple number line up to 20 outside in chalk. Then call out numbers for your child to jump to so they move forward and backwards over the line.

Place a sticker here.

Finding the matching number

Count the number of objects in each box. Circle the matching number.

How many frogs?

11 12 13

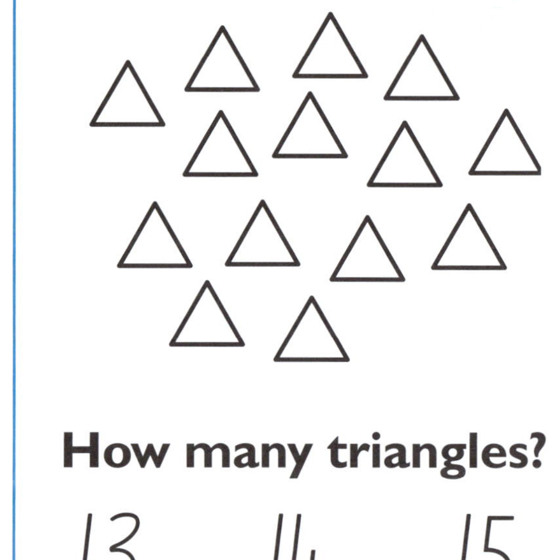

How many triangles?

13 14 15

How many spots?

14 15 16

How many hairs?

12 13 14

1. Make some simple number cards with one number on each, and ask your child to match the correct card to each picture in this activity.

Place a sticker here.

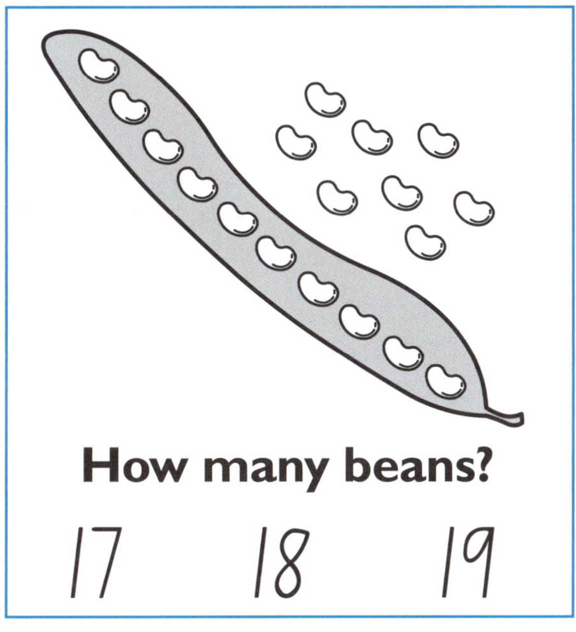

How many beans?

17 18 19

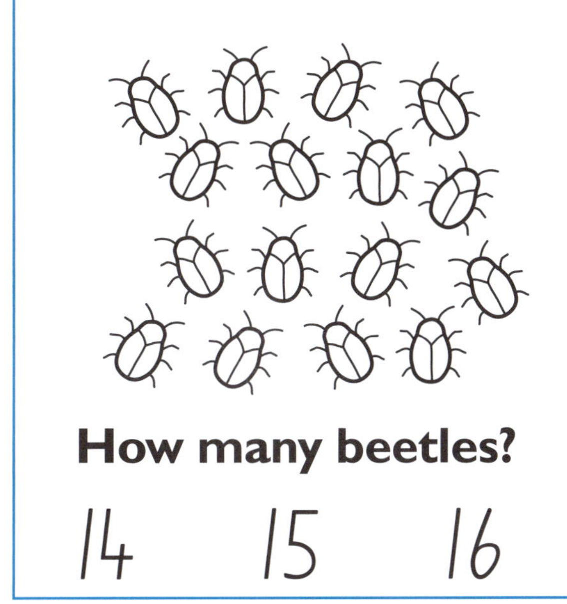

How many beetles?

14 15 16

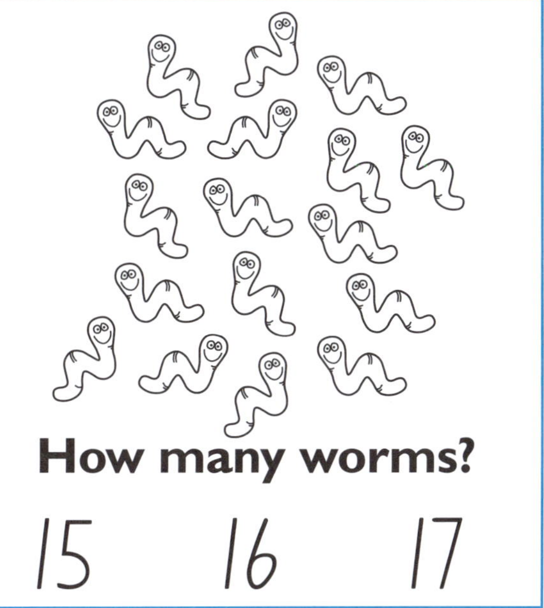

How many worms?

15 16 17

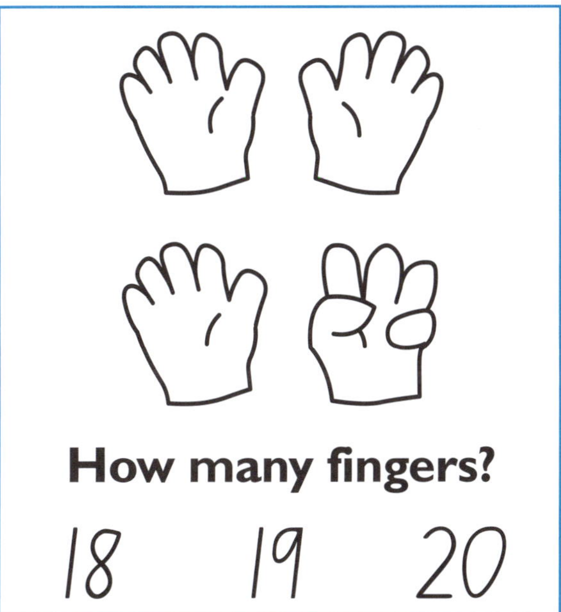

How many fingers?

18 19 20

2. Give your child 5 different dominoes. Ask them to count the dots, then match a number card to each domino.

Place a sticker here.

Writing numbers before and

Write the numbers that come before and after the numbers on the middle shirt.

13 14 15

16 17

11 12

1. Ask your child to close their eyes. Ask them what number comes before 15? After 17? Before 19?

Place a sticker here.

after from 1-20

18 19

17 18

15 16

2. Make some simple number cards up to 20 with one number on each. Jumble them up, then ask your child to put them in the correct order.

Place a sticker here.

Writing the missing numbers

Write the missing numbers on each snake to show how you count from 10 to 20.

1. When a snake is complete, ask your child to continue counting aloud as far as they can go.

Place a sticker here.

from 10-20

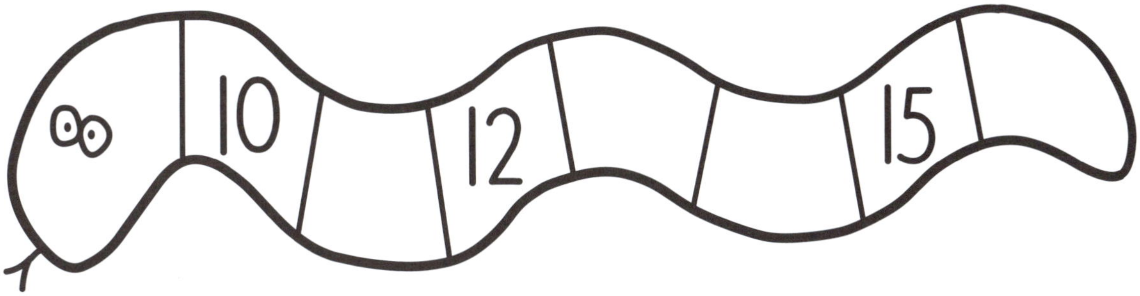

10
12
15

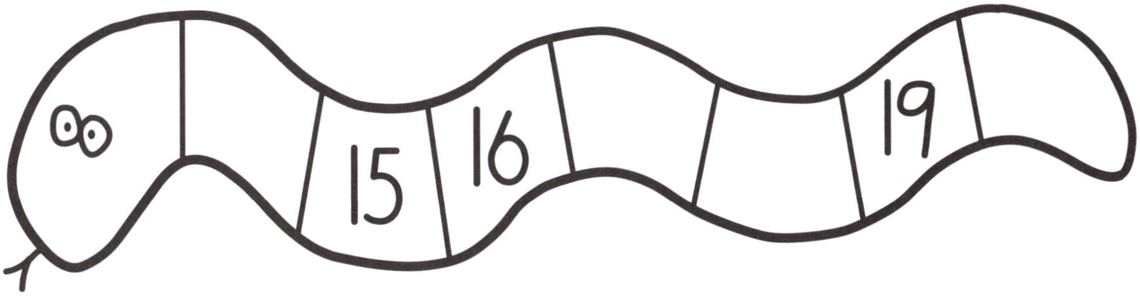

15
16
19

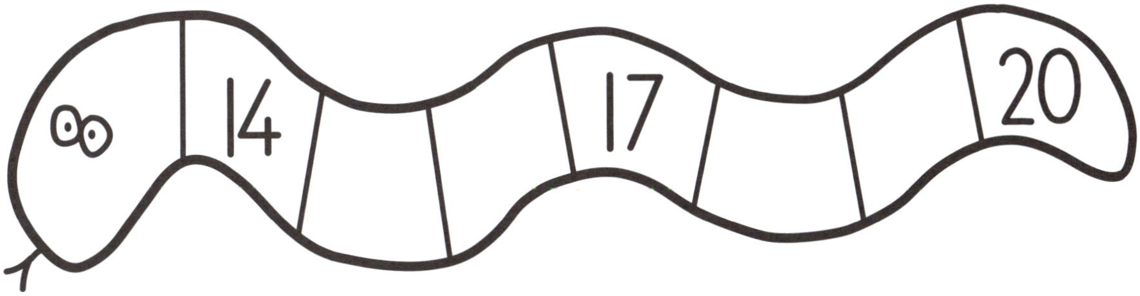

14
17
20

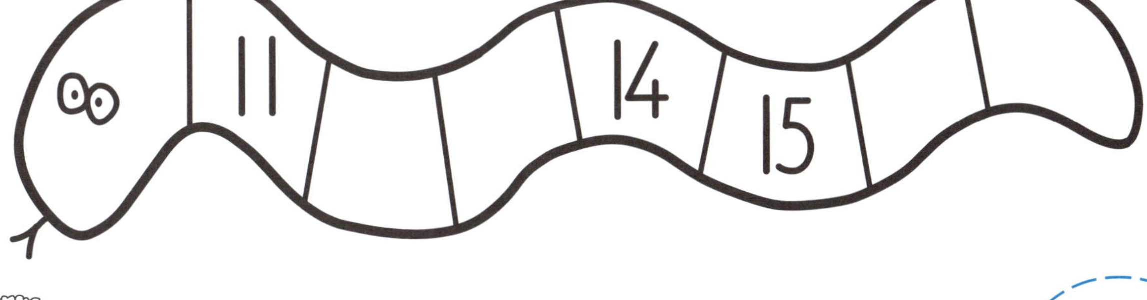

11
14
15

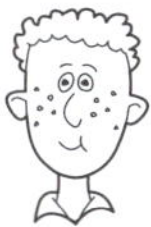

2. When a snake is complete, ask your child to start at the tail and count aloud backwards.

Place a sticker here.

Writing the missing numbers

Every second number is missing. Write the missing numbers to show how you count to 20.

1	2	3		5
	7		9	
11		13		15
	17		19	

1. Before your child fills in the boxes, ask them to count aloud all the numbers in black. This will introduce your child to the idea of 'odd' and 'even' numbers.

Place a sticker here.

	2		4	
6		8		10
	12		14	
16		18		20

2. Roll 3 dice. Ask your child to count the dots, then point to the numeral in the box on this page that matches the number.

Place a sticker here.

Joining the dots to 20

Here are 2 hidden pictures. What do you think they might be? Join the numbers from 0 to 20 to make the pictures. Start at 0 each time.

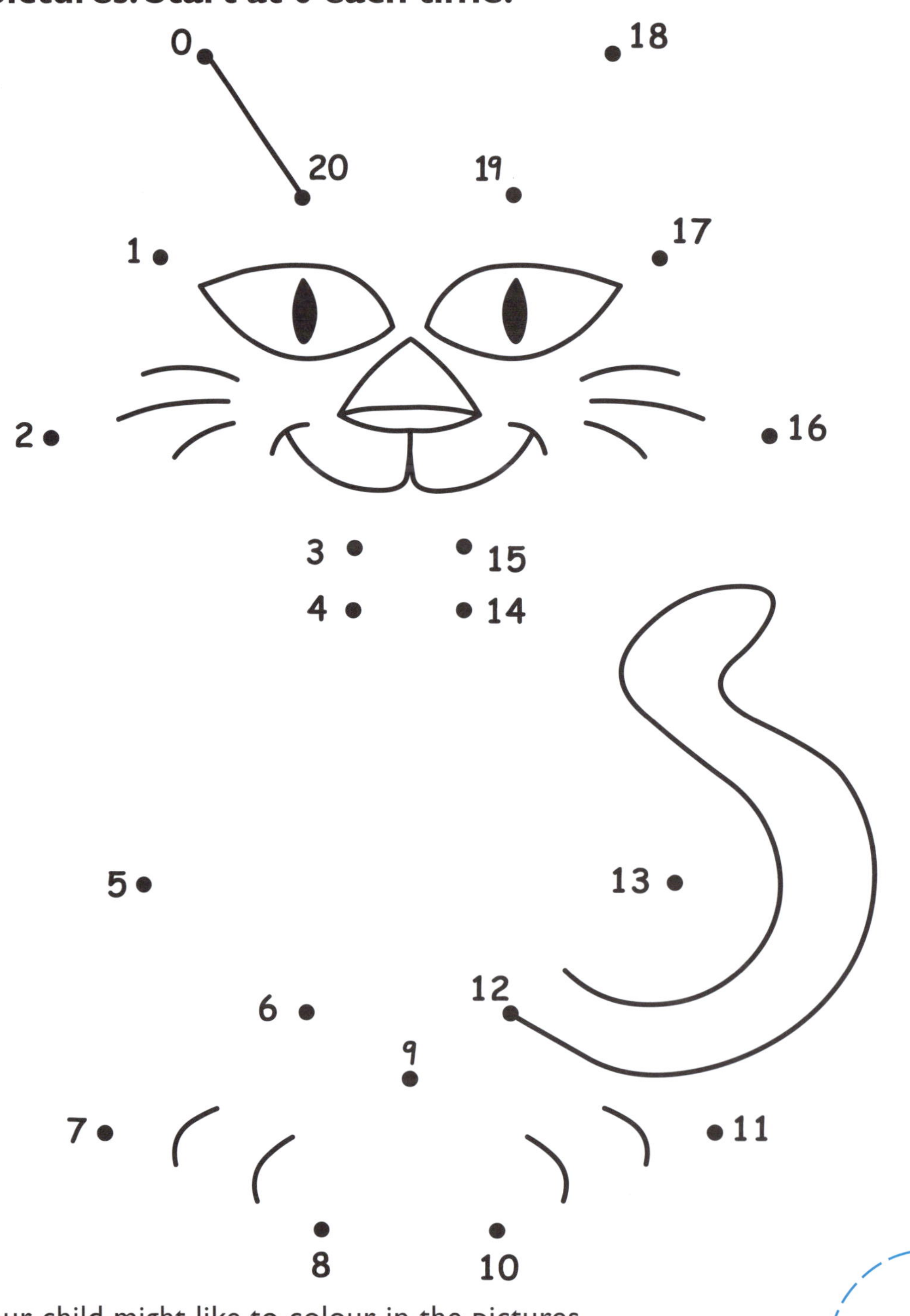

1. Your child might like to colour in the pictures.

Place a sticker here.

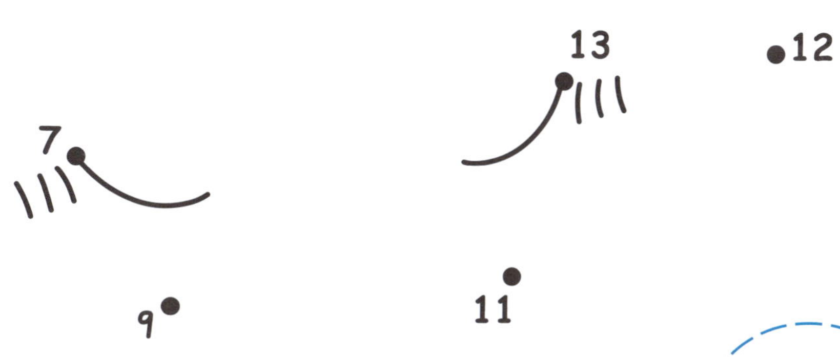

2. Help your child make their own dot-to-dot activities with simple shapes or pictures.

Place a sticker here.

Well done!

You have finished the book!

**Place your last two stickers on the picture.
You can colour in the picture, too.**